Making the Reserve Retirement System Similar to the Active System

Retention and Cost Estimates

Michael G. Mattock, Beth J. Asch, James Hosek

Prepared for the United States Army
Approved for public release; distribution unlimited

The research described in this report was sponsored by the United States Army under Contract No. W74V8H-06-C-0001.

Library of Congress Cataloging-in-Publication Data is available for this publication.

ISBN 978-0-8330-8418-7

The RAND Corporation is a nonprofit institution that helps improve policy and decisionmaking through research and analysis. RAND's publications do not necessarily reflect the opinions of its research clients and sponsors.

Support RAND—make a tax-deductible charitable contribution at www.rand.org/giving/contribute.html

RAND® is a registered trademark.

Preface

This document reports results from a research project titled "Analyses of Changes to Military Retirement." The purpose of the project was to identify potential costs or savings of changes to Army reserve component (RC) retirement that align it with that offered to active component (AC) soldiers. Of potential concern was whether this change would threaten the Army's ability to sustain RC strength, how it might affect flows of AC members to the RC, and whether the experience mix of the RC would change in a way that would pose challenges for Army manning. In this document, we present and discuss the results of that analysis. RAND researchers have developed a model of AC retention and RC participation, known as the dynamic retention model (DRM), that permits analysis of how the AC and RC strength and experience mix would change under alternative AC and RC compensation proposals. The DRM has been used to support military compensation decisions in the past, and the RAND Arroyo Center used it to conduct an analysis of the Army proposal. The findings should be of interest to the Army and others in the policy community concerned about military compensation and specifically compensation for RC members and, more broadly, to those with an interest in military manpower, personnel, and compensation issues.

This research was sponsored by MG John G. Rossi, director of the Army Quadrennial Defense Review Office within the Office of the Deputy Chief of Staff, G-8, and conducted within the RAND Arroyo Center's Manpower and Training Program. RAND Arroyo Center, part of the RAND Corporation, is a federally funded research and development center sponsored by the United States Army.

The Project Unique Identification Code (PUIC) for the project that produced this document is HQD136506.

Contents

Figures

Tables

Summary

RAND Arroyo Center analyzed how offering a retirement annuity immediately to vested reserve component (RC) members, both enlisted personnel and officers, would affect RC participation, as well as active component (AC) retention, both in the steady state and during the transition period. We also estimated the change in personnel costs in the steady state and considered how quickly RC participation and AC retention would change if currently serving RC members were given a choice between the two retirement systems. Because it describes the underlying decision process by which members choose to stay in or leave the AC or stay in or leave the RC, the dynamic retention model (DRM) was used to simulate and evaluate the retention effects of policies that have yet to occur. The model is designed to show the supply response to the policy change and therefore the effect on participation and retention if all who voluntarily choose to stay in the AC or to participate in the RC were permitted to do so. The advantage of modeling these voluntary behaviors is that we can identify the changes that could occur if they were allowed to do so, in the absence of other policy interventions. At the same time, personnel management actions could be designed, targeted, and implemented to limit or regulate the voluntary choices, consistently with the Army's desired force size and shape. We find that RC participation among those with prior AC service increases in the midcareer because of the greater value of RC retirement benefits when vested members can claim benefits immediately. For those with more than 20 years of AC and RC service, two effects operate. On the one hand, a member would have a stronger incentive to leave because he or she can claim RC retirement benefits immediately; on the other hand, a member could increase his or her retirement benefit by staying longer. For enlisted personnel, we conclude that the first effect dominates, and we find that RC participation would drop markedly after 20 years of service (YOS). For officers, the empirical results imply that the second effect dominates initially and that RC participation actually would rise in the first few years beyond 20 YOS. After that, RC participation also would fall for officers. For RC members with no prior AC service, we find that RC participation would be essentially unchanged; these RC members could typically expect small retirement benefits, so the changes in incentives produced by the policy change would be minor. Overall, the RC force would decrease in size, become more junior, and leave the Army with less seniority.

We find that the policy would have small but discernible effects on AC retention. AC retention would increase in midcareer (that is, from YOS 10 to 20) but decrease near and immediately after retirement vesting. AC retention would increase in midcareer because AC and RC service would become more valuable as a result of the policy change. AC retention would decrease near and immediately after retirement vesting because members who have particularly good civilian opportunities would have little to lose by leaving, even close to the vesting point.

In the steady state, Army personnel costs would fall. This would occur in large part because the AC force would become more junior, more AC members separate prior to YOS 20, and those who do make it to 20 tend to leave the Army with less seniority. Decreases in RC costs, which come from changes in personnel costs of enlisted personnel with prior AC experience, would also contribute to the cost savings, but most of the cost savings would be due to changes in the AC force. Also, if AC force is held constant, the potential change in retirement benefits alone would lead to a decrease in RC force size. However, continuation pay could be offered to sustain the RC force, and, although this would add expense, total cost savings would still be $790 million annually in the steady state.

In the transition to the steady state, the speed at which the change in participation would occur varies with the transition policy. If the Army were to place currently serving RC members in the new system but give them the option to stay in the current system, the change would occur more quickly than if currently serving RC members were kept under the current system and not given any choice. We find that much, though not all, of the change in participation would occur within the first five years after implementation of the policy, when members have a five-year enrollment period. Changes to AC retention also would occur more quickly when members have this choice. The rapid speed is not surprising given that we find that relatively few would prefer the current system and that those who do would tend to be significantly older and near the current retirement eligibility age of 60. (Although our analysis does not estimate costs during the transition phase, other studies indicate that cost savings would occur more rapidly when members are allowed to choose between new and old systems.) Given that most RC members would choose the new system, we expect that cost savings would be realized quickly if RC members had a choice of systems. Furthermore, because members voluntarily choose whether to switch, members would be made better off by having the choice; they would select the alternative that is best for them. Thus, providing members with an immediate option to transition to the new system has the potential to make service members better off *and* lower costs.

Acknowledgments

We are grateful to Timothy Muchmore, Army Quadrennial Defense Review Office within the Office of the Deputy Chief of Staff, G-8, for suggesting this topic and for his guidance and support throughout the research. We thank John Warner and our RAND colleagues Michael L. Hansen and Jennie W. Wenger for their helpful review of this work.

Abbreviations

AC	active component
BAH	basic allowance for housing
DRM	dynamic retention model
FY	fiscal year
NPS	non–prior service
QRMC	Quadrennial Review of Military Compensation
RC	reserve component
WEX	Work Experience File
YOS	year of service

Chapter One. Introduction

Should Active and Reserve Retirement Benefits Be the Same?

Today's reserve component (RC) forces play both strategic and operational roles. Members of the Selected Reserves, in particular, have deployed extensively throughout the military operations in Iraq and Afghanistan and will no doubt deploy in future contingencies. To help ensure that the RC sustains its operational capability and, at least over the next five to ten years, retains reservists who have gained experience in the past decade of deployments, reservists must receive adequate compensation to secure their participation. They should perceive their compensation as being equitable with respect to active component (AC) compensation while recognizing that reserve service is part time during periods of nonactivation.

Retirement benefits are a major element of AC compensation and an important element of reserve compensation. However, the two systems differ, as we describe in more detail in Chapter Two. Retirement benefits start immediately upon retirement for AC personnel with at least 20 years of service (YOS) but do not start until age 60 for reservists (or somewhat sooner, depending on the extent of a reservist's deployment) with at least 20 creditable YOS. An AC member with 15 YOS could receive an immediate, lifetime stream of retirement benefits upon completing five more years in the AC; if that member left the AC and participated in the reserves for five more years, retirement benefits would not start until age 60. The difference in active and reserve retirement benefit systems may seem inequitable to service members—for both reservists and AC members thinking of joining the reserves. Further, the difference in these systems could become more of an issue if reservists continue to be deployed regularly as part of the total force, rather than being conserved as a strategic asset. If so, the apparent inequity in the systems might hurt morale, decrease reserve supply, and reduce readiness.

As an alternative to the current system, the RC retirement system could be reshaped to be the same as the AC system in specific ways described below. But would it be worth doing this?

The U.S. Army Asked the RAND Arroyo Center to Assess the Effects of a Retirement Reform Proposal

The Army requested that we address the question of the impact on the Army of reshaping the RC retirement system to be like the AC one. The Army wanted to know how the proposal would affect RC participation, AC retention, the flow of members from the AC to the RC, and personnel costs. Would the new policy enable the RC to sustain its current force size? Would there be changes in the experience mix? Would the flow from the AC to the RC change? Would there be repercussions for the AC force? Intuitively, the immediate availability of RC retirement benefits increases the generosity of the RC retirement system and increases the incentive to participate in the reserves. However, the decrease in retirement points decreases the generosity of the system, and analysis is required to see how these effects interact. If the proposal makes the RC more attractive, the incentive to move from the AC to the RC could increase, creating downward pressure on AC retention and strength.

Also, immediately available RC retirement benefits could change the RC experience mix. The proposal changes the incentive to participate in the RC differently before versus after 20 YOS. Before 20 YOS, the incentive to participate increases relative to the current system because the payoff of reaching 20 years is higher. But after 20 years, the incentive to participate decreases relative to the current system because reservists can retire with immediate retirement benefits.

Another question is whether the proposed system would cost more or less than the current system, where cost includes not only retirement benefits but also supplemental current pay if needed to sustain a desired size and shape of the RC force. Related to this question is how the proposal would change the size and shape of the RC force relative to its *current* shape.

In assessing the RC retirement proposal, the Army asked us to assume that AC and RC retirement systems would continue to vest at 20 YOS. However, in a major change, RC members, like AC members, would begin receiving retirement benefits immediately upon retirement after 20 or more YOS. A reservist would receive a retirement point for each day of service; there would no longer be a point for each drill. This would result in 38 points per year if the reservist were not activated, and more if activated. Participation points—15 per year—would be eliminated.

The purpose of this study is to address these questions, making use of a model RAND researchers have developed, called the dynamic retention model (DRM), of AC and RC retention that permits simulations of new and untried policies. The focus of the analysis is on how the Army proposal would affect the size and shape of the AC and RC Army forces, as well as cost, relative to their current size, shape, and cost. We note that the approach we use could also be used to determine the pay changes and cost of sustaining other force sizes and shapes, though we do not pursue that question in this report. Because we have described the model extensively in past studies, we do not give a detailed description here. Instead, we provide a brief overview and then turn to the main results of the analysis in Chapter Two.

Overview of Our Approach

RAND's DRM is well suited to the analysis of structural changes in military compensation, such as the proposed RC retirement system. Recent applications of the model include analyses for the ninth, tenth, and 11th Quadrennial Reviews of Military Compensation (QRMCs) and analyses of other proposals for reserve retirement reform. The model's capability has steadily increased; for instance, new, faster estimation and simulation programs have been written, costing has been refined, and the model can now show retention and cost effects in both the steady state and the year-by-year transition to the steady state.

The model is based on a mathematical model of individual decisionmaking over the life cycle in a world with uncertainty, and the parameters of this model are empirically estimated with data on military careers drawn from administrative data files. The main version of the model begins with service in the AC, and individuals make a stay/leave decision in each year. Those who leave the AC take a civilian job and, at the same time, choose whether to participate in the RC. The decision of whether to participate in the RC is made in each year, and the individual can move into or out of the RC from year to year. More specifically, a reservist can choose to remain in the RC or to leave it to be a civilian, and a civilian can choose to enter the RC or remain a civilian. Another version of the model focuses on individuals who join the RC with no prior AC service. Both versions have been used in the analysis.

The data we use is the Defense Manpower Data Center Work Experience File (WEX). The WEX contains person-specific longitudinal records of active and reserve service. We use the WEX data for service members who began their military service in 1990 or 1991 and track their individual careers in the AC and, if they join, the RC, through 2010, providing 21 years of data on 1990 entrants and 20 years on 1991 entrants. For each AC component, we drew samples of 25,000 individuals who entered the component in fiscal year (FY) 1990–1991, constructed each service member's history of AC and RC participation, and used these records in estimating the model. We supplement these data with information on active, reserve, and civilian pay. AC pay, RC pay, and civilian pay are averages based on the individual's years of AC, RC, and total experience, respectively. We use 2007 military pay tables, but, because military pay tables have been fairly stable over time, with few changes to their structure,[1] we do not expect our results to be sensitive to the choice of year. For civilian pay opportunities for enlisted personnel, we use the 2007 median wage for full-time male workers with associate degrees. For officers, we use the 2007 median wage for full-time male workers with bachelor's degrees or more. The data are from the U.S. Census Bureau.

A limitation of the current state of this modeling is that the Army National Guard and Army Reserve are not treated separately but are combined into a single group, the Army RC. The model assumes that military pay, promotion policy, and civilian pay are time stationary, and we exclude demographics, such as gender, marriage, and spouse employment, from our model. We also exclude health status and health care benefits, and we do not explicitly model deployment or deployment-related pays. That said, the estimated models fit the observed data extremely well for the both the AC and the RC.

---

[1] An exception was the structural adjustment to the basic pay table in FY 2000 that gave larger increases to midcareer personnel who had reached their pay grades relatively quickly (after fewer YOS). A second exception was the expansion of the basic allowance for housing (BAH), which increased in real value between FY 2000 and FY 2005.

The approach is documented in several RAND reports (for instance, Mattock, Hosek, and Asch [2012], *Reserve Participation and Cost Under a New Approach to Reserve Compensation*, a technical report prepared for the 11th QRMC; and Asch, Hosek, and Mattock [2013], *A Policy Analysis of Reserve Retirement Reform*, a report prepared for the Office of the Assistant Secretary of Defense for Reserve Affairs).[2]

Organization of This Document

The remainder of this document presents the core findings of the analysis. We present steady-state results, as well as transition results. First, in Chapter Two, we discuss the steady-state retention and cost effects of the proposal. This part of the analysis begins by showing the impact of the proposal on RC participation in the steady state for individuals who had prior service in the Army AC, and this is followed by results showing the proposal's impact on Army AC retention. We also show the impact on RC participation for individuals who had no prior Army AC or RC service when they joined the Army RC. In Chapter Three, we discuss the second type of results, which are for the impact of the proposal during the transition period. The transition period is the time until a new 20-year-old AC entrant reaches age 60, or 40 years. We present two variants of the transition analysis: (1) future military entrants are under the new system while current AC and RC members are grandfathered in the current system (also referred to as the old system); and (2) current members are under the new system unless, in the year the new system is implemented, they opt to stay under the old system, and new RC members (specifically, AC members and civilians who choose to enter the RC within a five-year window from the start of the new system) are also under the new system unless they opt to stay under the old system. Chapter Four offers a brief conclusion, and an appendix provides simulation tables.

---

[2] Goldberg (2001) provides an extensive discussion of the history of retention models. Gotz (1990) provides a detailed discussion of the advantages of the DRM approach relative to other approaches that have been used to assess the effects of compensation proposals on retention.

Chapter Two. Steady-State Retention and Cost Results

We discuss the steady-state results in this chapter. These results show the RC and AC retention profiles that result after the new RC retirement system is fully mature, and Chapter Three discusses retention during the system's transition to maturity. The results therefore provide information on the force and shape that could emerge in the long run. This chapter also presents the RC and AC steady-state costs that would emerge under the proposal versus the current system. Before presenting the results, we provide a brief overview of the reserve compensation system.

A Brief Overview of the Reserve Compensation System

RC compensation differs from AC compensation in several respects. A detailed review is provided in Mattock, Hosek, and Asch (2012) and Asch, Hosek, and Mattock (2013). In brief, different types of RC duty status result in different earnings of basic pay plus allowances. All duty in the AC is active duty, and the member receives one day of basic pay for a day of active duty. In the RC, duty may be inactive or active. When reserve units drill twice per day on inactive duty for one weekend (two days) per month, the reservist receives a day of basic pay for each drill—or two days of basic pay for one day of drill. When on active duty, the reservist is paid one day of basic pay each day, just like an AC member. Thus, RC members receive up to 48 "days" of drill pay for 24 days of inactive-duty training and 14 days of pay for annual training. While on annual training, members receive a housing allowance based on the schedule that applies to RC members and the basic allowance for subsistence.

AC and RC members also differ in how they accumulate service toward retirement benefits. Service in the RC is prorated depending on the number of drills (inactive duty) and days on active duty during a "year" of 360 days. For a year to be creditable toward reserve retirement vesting, an individual must accumulate at least 50 points. A drill earns the member one point, and a day of active duty earns one point. For example, a reservist who participates in four drills per month (one weekend with two drills per day) and 14 days of inactive-duty training in the summer accumulates (48 + 14 =) 62 points. Because an RC member also receives an additional 15 points for participating in the Selected Reserves, his or her total would be 77 points. YOS in the AC before joining the RC are also creditable. Following this example, a reservist with five years in the AC and 15 creditable years of inactive duty in the RC would have

$$\left(5 + \frac{15 \times 77}{360}\right)$$ YOS for the purpose of computing retirement benefits. The formula for computing

retirement benefits is the same for the RC and the AC:

$0.025 \times$ high-three basic pay $+$ effective YOS. (High-three basic pay is the average of basic pay over the highest 36 months of basic pay received by the member.)

Also, RC housing allowance rates differ from those of the AC. When in inactive-duty training, usually for two weeks in the summer, an RC member receives a partial allowance that does not vary across locations, unlike that of his or her AC counterparts. When the RC member serves on active duty for more than 30 days, his or her housing allowance is the same as that of his or her AC counterparts. When an RC member does weekend drills, he or she receives no housing allowance; each finds available space in government quarters.

Finally, as mentioned in Chapter One and of particular interest in terms of the policy we analyze is the fact that vested RC members generally do not receive retirement benefits until age 60, while vested AC members can receive benefits as soon as they separate. Both RC and AC members vest after completion of 20 YOS. At the request of the Army, we analyze a policy alternative that would allow RC members to receive retirement benefits immediately, rather than wait to age 60, like their AC counterparts.

To help make concrete the nature of the proposal we analyze, we provide some comparisons of the annuity under the current versus the proposed change, as well as the present value to the individual. In the RC retirement proposal under consideration here, AC and RC retirement systems would continue to vest at 20 YOS. However, in a major change, RC members, like AC members, would begin receiving retirement benefits immediately upon retirement after 20 or more YOS. A reservist would receive a retirement point for each day of service; there would no longer be a point for each drill. This would result in 38 points per year if the reservist were not activated, and more if activated. Participation points—15 per year—would be eliminated.

Table 2.1 shows how this proposal would alter the monthly retirement benefit of a staff sergeant (E-6) and a lieutenant colonel (O-5) retiring with 20 YOS.[3] The rows of the table are for different combinations of active and reserve years, ranging from zero active years and 20 reserve years to 20 active years and zero reserve years, in steps of four years. The benefit computation is based on the 2013 basic pay table and uses the pay rates of an E-6 and O-5 with 18 to 20 YOS, $3,650.70 and $8,118.00, respectively. As seen, the monthly benefit is lower for any member with reserve service, and the decrease is greater the more years of reserve service. However, the proposal would start benefit payments immediately upon retirement rather than at age 60 or so. For a reservist retiring at age 45, this is 15 more years of benefits. This leads to a higher present value for the stream of payments, as in shown in Table 2.2. Of course, these tables of benefits do not take into account the probability of reaching 20 YOS. But our modeling approach accounts for this element and more, as noted earlier.

---

[3] E-6 and O-5 are typical ranks for enlisted and officers at 20 YOS.

Table 2.1. Monthly Reserve Retirement Benefit for a Staff Sergeant (E-6) and a Lieutenant Colonel (O-5) Under the Current and Proposed Systems

Years		E-6		O-5	
Active	Reserve	Current	Proposed	Current	Proposed
0	20	390	193	868	428
4	16	677	519	1,506	1,155
8	12	964	846	2,145	1,881
12	8	1,251	1,172	2,783	2,607
16	4	1,538	1,499	3,421	3,333

NOTE: Years are in reverse order in the reserve column because the active and reserve years are allocated to sum to 20 in each row.

Table 2.2. Present Value of Lifetime Reserve Retirement Benefit for a Staff Sergeant (E-6) and a Lieutenant Colonel (O-5) Completing 20 Active Plus Reserve Years of Service at Age 45 Under the Current and Proposed Systems

Years		E-6		O-5	
Active	Reserve	Current	Proposed	Current	Proposed
0	20	5,892	19,436	59,302	79,638
4	16	10,222	52,374	102,892	214,603
8	12	14,553	85,313	146,483	349,568
12	8	18,884	118,251	190,073	484,533
16	4	23,214	151,189	233,664	619,498

NOTE: Years are in reverse order in the reserve column because the active and reserve years are allocated to sum to 20 in each row. The present values in the table were computed using personal discount rates of 12.0 percent for Army enlisted personnel and 5.4 percent for officers. The discount rates were estimated as parameters of the DRM.

The Army wanted to know how the proposal would affect RC participation, AC retention, the flow of members from the AC to the RC, and personnel costs. Would the new policy enable the RC to sustain its current force size? Would there be changes in the experience mix? Would the flow from the AC to the RC change? Would there be repercussions on the AC force? Intuitively, the immediate availability of RC retirement benefits increases the generosity of the RC retirement system and increases the incentive to participate in the reserves. However, other things equal, the decrease in retirement points decreases the generosity of the system. Still, as Table 2.2 makes clear, the new policy would increase the present value of reserve retirement benefits relative to the current policy, which increases the incentive to participate in the reserves. For someone service in an AC who is contemplating participation in the reserves after leaving the AC, the higher present value of reserve retirement benefits increases the value of continuing in the actives. But at the same time, it increases the value of leaving the AC and participating in the reserves. Analysis is required to see how these effects interact. Over what YOS would it make sense to continue in the actives, and when would it make sense to shift to the reserves? Also, for individuals who leave the actives and become civilians, how would the increased generosity of reserve retirement benefits affect their participation in the reserves? If individuals shift from the AC to the RC, how much downward pressure would there be on AC retention and strength? The remainder of this section and the next presents the results of our analysis.

Participation Among Reserve Component Members with Prior Active Component Service

Figure 2.1 shows Army RC enlisted and officer participation by YOS in the steady state under the current RC retirement system (black line) and the proposed system (red line). The underlying population is soldiers who first served in the AC. These lines represent overall participation behavior as computed from large-scale simulations of individual decisionmaking behavior under the current and proposed policies. The simulations employ estimates of the model's parameters based on actual AC and RC retention data and described in the reports cited earlier.

The RC retirement proposal affects both overall participation—RC end strength—and the shape of participation by YOS. As seen, the proposal produces higher participation between ten and 20 YOS for both enlisted personnel and officers and lower senior participation for enlisted personnel, as compared with the current system. For officers, senior participation increases until YOS 25. Overall enlisted participation decreases by 6 percent (from 178,491 to 167,882—see Figure 2.1), while officer participation increases by 5 percent (from 18,617 to 19,541). In the case of enlisted personnel, the drop in participation comes almost entirely from the post–20 YOS range. Enlisted RC strength can be restored through the use of RC continuation pay. We discuss later in this chapter the implications for participation and cost savings of adding RC continuation pay.

Figure 2.1. Simulated Steady-State Effect of the Proposal on Army Enlisted and Officer Reserve Component Participation, Among Those with Prior Active Component Experience

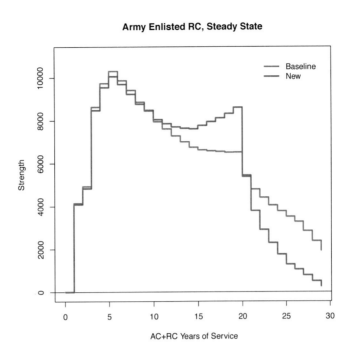

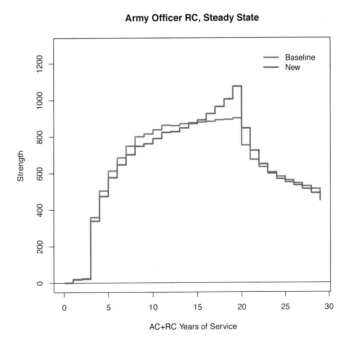

The most profound effect of the new system is the change in RC participation by YOS. For enlisted soldiers with ten YOS and officers with 12 years, where YOS count both AC and RC years, participation is slightly higher than at baseline, and the gain in participation grows larger up to 20 YOS. This gain is driven by the increasingly high value placed on the reserve retirement benefit stream. As YOS climb toward 20 years, benefits are discounted over fewer years, and their value increases nonlinearly, spurring participation in the RC. When 20 years have been reached, benefits can begin immediately. This strengthens the incentive to leave on the one hand but also increases the value to staying on the other, because members who stay also accumulate more points and therefore higher retirement benefits. For enlisted personnel, the stronger incentive to leave at 20 offsets the additional value to staying longer. Thus, the proposal creates strong incentives to complete 20 YOS for enlisted RC members with ten or more YOS, and strong incentives to leave after completing 20 years. For officers, the additional value of staying offsets the additional incentive to leave at 20. Officers have a lower personal discount rate and so place a higher value on future benefits than do enlisted personnel. Consequently, officer RC participation actually increases after 20 years, up to year 25, despite the availability of immediate retirement benefits.[4]

Part of the change shown in the figures reflects the change in the number of AC leavers who ever participate in the RC, from 36 percent to 38 percent among enlisted and from 27 percent to 30 percent among officers.[5] As we discuss later, to some extent, this reflects more members leaving the AC between 15 and 19 YOS and participating in the RC. However, most of the increase in RC participation in YOS 10 to 20 results from an increase in overall RC participation among members leaving the AC before YOS 15. Tied to this is the fact that people who previously stayed for much longer than 20 years are choosing shorter careers, as shown by the decrease in RC participation after YOS 20.

Whether such a change in RC enlisted and officer forces is desirable is a matter for Army leadership to consider. The change in experience mix translates into a younger RC force with prior AC experience: The average age of the RC force with prior AC service falls by 2.4 years among enlisted personnel and by 1.6 years for officers. Notice also that RC participation is slightly lower in years 3 to 10 for enlisted and in years less than 12 for officers. At the margin,

---

[4] The personal discount factor is a parameter we estimate in our model. We estimate the discount factor for each service, for officers and for enlisted personnel. For Army enlisted, we estimate a personal discount factor of 0.893. For Army officers, we estimate a personal discount factor of 0.949. These are equivalent to personal discount rates of 12.0 percent and 5.4 percent respectively, using the relationship $\beta = \dfrac{1}{1+r}$, where β is the discount factor and r is the discount rate. For a further discussion of our estimates, see Mattock, Hosek, and Asch (2012).

[5] These are the percentages based on our policy simulations. Other research has shown that RC participation rates can be affected by RC affiliation bonuses (Hosek and Miller, 2011). That study also describes the increase in RC bonus generosity and usage during the past decade. Thus, RC bonuses are a tool that could be used in the future to sustain or increase the AC-to-RC join rates.

AC members have a stronger incentive under the new retirement system to serve longer in the AC and then join the RC in midcareer years.[6]

The effect of the proposal on RC participation by age is more easily seen in Figure 2.2. The figure shows RC participation by age, rather than by years of active and reserve service for enlisted personnel and officers. Under the baseline, the number of RC enlisted participants among those with prior AC service is fairly constant between ages 25 and 55, declining between ages 55 and 60. Under the proposed alternative, enlisted RC participation increases between ages 30 and 40 and declines thereafter. A broadly similar pattern is observed for officers. RC participation for officers increases with age under the baseline, until just over age 50, and then declines. Under the proposed alternative, participation increases among younger personnel between ages 30 and 42, with a particularly large increase at age 42, and then decreases after age 42, and especially after age 50.

---

[6] As mentioned, a limitation of the study is that it does not consider deployment. However, one could speculate that the greater incentive to participate in the RC up to YOS 20 could help to support and stabilize reserve unit manning for deploying reserve units.

Figure 2.2. Simulated Steady-State Effect of the Proposal on Army Enlisted and Officer Reserve Component Participation by Age, Among Those with Prior Active Component Experience

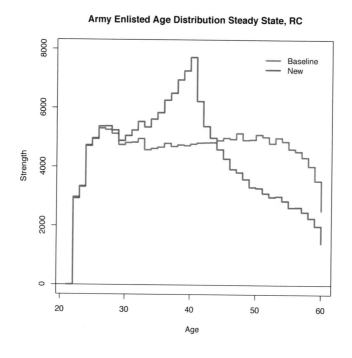

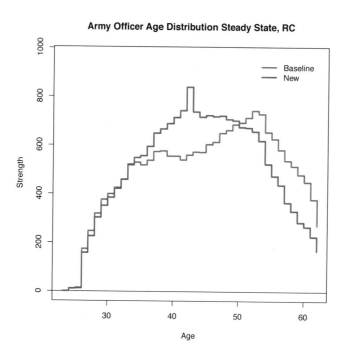

It is important to recognize that the model is designed to show supply behavior, i.e., an individual's willingness to participate in the RC. Figure 2.1 can therefore be interpreted as showing the retention profile that could materialize under a passive personnel policy that allowed the participation of all individuals willing to participate, i.e., there would be RC billets available for all who are willing to participate. However, whether the change shown in the figure would materialize will depend on Army RC policy, laws, and regulations concerning available billets.

There are at least two aspects to consider. First, the model assumes that growth in average RC pay by YOS will be the same under the proposal as it is under the current system. However, the increase in the number of midcareer personnel could clog promotion channels, which would decrease RC pay at higher ranks. If that occurred, the impact of the proposal on midcareer RC participation would be smaller than shown. The negative effect of slower promotion could also be felt at higher ranks in post-20 years, and that would further decrease the incentive to stay (strengthen the incentive to leave). On the other hand, post-20 RC members who enjoy their work and believe that they are still competitive for promotion and increased responsibilities may opt to continue to participate in the post-20 years. Indeed, we find an increase in post-20 participation among those with prior AC experience among officers.

Second, the RC might restructure in a way that created more slots for midcareer personnel and fewer slots for senior personnel. If this occurred, junior and early midcareer RC personnel would not have to worry about lower promotion rates to midcareer ranks. But to the extent that opportunities for promotion to senior ranks decreased, there would be a decrease in expected future RC pay, and this would have some negative effect on midcareer RC participation— shifting the new participation curve below the red line shown but most likely well above the black line because of the immediate availability of retirement benefits upon reaching 20 years.

Finally, the RC could become more selective when making midcareer promotions. This would mean that fewer of those willing to participate in the RC would be selected for promotion. Increasing the selectivity for promotion is different from passively allowing promotion channels to clog. In the latter, individuals self-select on whether to stay, and those choosing to stay might have the strongest affinity (taste) for reserve service but will not necessarily be the best performers. In contrast, an assertive policy of more-stringent standards for promotion would—if well implemented—have the potential of increasing the quality of midcareer personnel and increasing the quality of the pool of those promotable to the highest ranks. Related to this point, it might be the case that higher performers also have stronger tastes for RC service. If so, they would be more likely to stay after 20 years (the post-20 participation line would shift up a bit).

Active Component Retention

Figure 2.3 shows the effect of the proposal on AC retention. Changes in RC compensation can affect AC retention: By making benefits payable immediately upon retirement, the proposal makes RC participation more valuable to AC soldiers in the junior ranks. AC retention in the junior ranks shows an increase, with the effect more pronounced among officers. This increase in junior rank retention is due to an increase in the value of a military career overall; the ability to leave the AC with the option of joining the RC and being able to collect an RC retirement benefit that has the same payoff per AC YOS as under the AC retirement system.

Figure 2.3. Simulated Steady-State Effect of the Proposal on Army Enlisted and Officer Active Component Retention

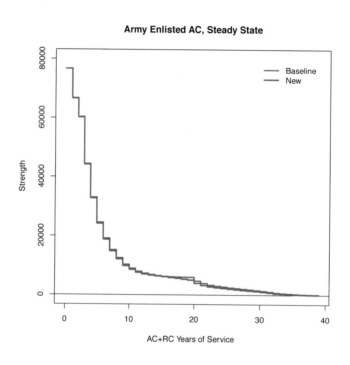

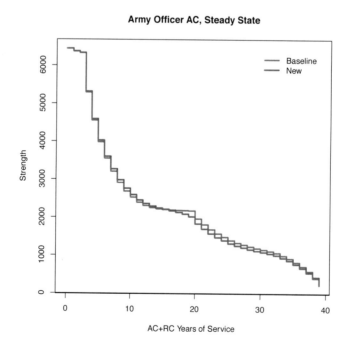

But AC retention is also lower in years 18, 19, and 20. The future option to join the RC that helped to pull junior members forward is exercised by some members in these years. Intuitively, we can see that an AC soldier who is nearing 20 YOS will give up very little in retirement benefits by switching to the RC, accepting a possibly high-paying civilian job offer, and leaving the RC soon after reaching 20 creditable years. The high-paying civilian job offer is only an example but is realistic in the sense that a soldier with almost 20 years in the AC who receives an attractive external opportunity or perhaps has family issues requiring his or her time and attention is most likely to follow this route. That outflow decreases the number of AC soldiers at years 18, 19, 20, and, because of this, the number of personnel beyond 20 is also lower than at baseline. Still, on the whole, the impact of the proposal on AC retention is small.

We can quantify the fraction of RC accessions with prior AC service who join with 15 to 19 YOS. The percentage increase among officers is from 2 percent to 12 percent, and from 1 percent to 5 percent for enlisted personnel. Thus, a higher percentage of people who join the reserves are coming from the midcareer AC.

Participation Among Non–Prior-Service Reserve Component Members

As Figure 2.4 shows, the proposal has a positive effect on RC enlisted participation among reservists with no prior service, but the effect is small. Retirement benefits, being available sooner (after 20 years rather than at age 60), increase the perceived value of an RC career and cause an increase in RC participation in all years up to 20. After 20 years, there is a slight decrease in participation. We find no change in non–prior-service (NPS) RC participation among officers.

The proposal's effect on NPS RC participation differs markedly from the effect on RC participation for soldiers with prior AC service. The previous discussion about incentives to stay until 20 and to leave after 20 still applies, but the difference is that NPS RC participants typically can expect small retirement benefits. At an extreme, if a NPS reservist were never activated and accumulated just 38 points per year, the reservist would have only $\left(\dfrac{20 \times 38}{360}\right) = 2.1$ effective YOS. For a given pay grade at 20 YOS, this reservist's retirement benefits would be only one-tenth of those for an AC soldier completing 20 years and about one-fourth of those for a reservist with five years in the AC and 15 in the RC, who would have 8.2 effective years. If the reservist received them at age 45, say, instead of age 60, the proposal would offer 15 more years of benefits, but the benefits would be about half as large as today's benefits. This is because the reservist would accumulate 38 points per year, down from 77 points per year, in this simple example. The simulation indicates that the benefit amount under the proposal, although available immediately upon retirement, is not large enough to cause much change in participation among the NPS RC participants.

Figure 2.4. Simulated Effect of the Proposal on Army Enlisted and Officer Non-Prior-Service Reserve Component Participation

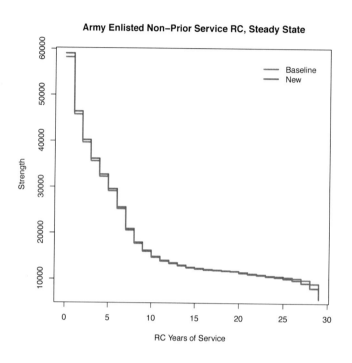

Army Enlisted Non-Prior Service RC, Steady State

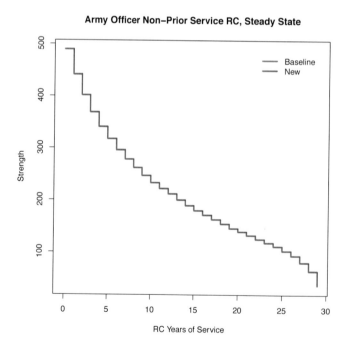

Army Officer Non-Prior Service RC, Steady State

Figure 2.5 provides a summary of the effects on participation for enlisted RC personnel. The figure shows total participation of enlisted RC members, including those with and without prior AC experience, both in the baseline and under the new proposal. (For members with no prior AC experience, only reserve years count toward YOS; for members with prior AC experience, both active and reserve years count.) The proposal can be expected to change the experience mix of reservists. It increases the number of midcareer personnel serving until 20 years and decreases the number of senior personnel. In the case of officers (not shown), it increases the number of midcareer personnel and those with more than 20 years. For enlisted personnel, the RC experience profile would look more similar to the one for AC personnel, whereby relatively few members serve for longer than a 20-year career. That said, whether the change shown in the figure will actually occur would depend on Army RC policy, laws, and regulations concerning available billets.

Figure 2.5. Simulated Effect of the Proposal on the Total Army Reserve Component Enlisted Force (Prior Active Component and Non–Prior Active Component Service)

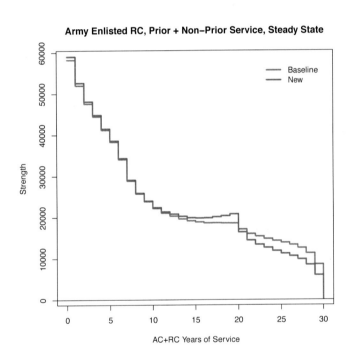

Army Personnel Costs

Table 2.3 shows estimated personnel costs and how costs change under the proposal. The cost estimates are based on the simulation results for the steady state under the current and proposed RC retirement systems. Costing is computed for three groups: AC soldiers, RC soldiers with prior AC service, and RC soldiers with no prior service. For each group, we compute cash compensation costs (those that make up regular military compensation in our analysis) and retirement costs. The also shows force size at baseline and under the proposal. The grand totals are shown at the bottom of the table. Overall, we find that Army personnel costs decrease under this proposal because of the change in the experience mix of the AC and, specifically, because the AC becomes more junior and AC members leave the Army with less seniority.

As the table shows, by construction, there is no change in AC force size; a decrease of 10,600 or 6 percent in prior-service RC enlisted members and an increase of 924 officers or 5 percent in prior-service RC officers; and a decrease of 800 enlisted members or 1.5 percent in NPS RC force size and no change in officer NPS RC force size. As mentioned, there is a slight decrease in prior-service RC participation in years 3 through 10 for enlisted and a somewhat larger decrease in RC participation in years prior to year 12 for officers, which is mirrored by a corresponding increase in AC retention. Also, there is a decrease in the number of AC personnel in years 18, 19, 20, and beyond as the AC force becomes more junior overall. This decreases the current cost of AC personnel. In addition, the decrease in force size after year 20 causes a decrease in AC retirement cost for both officers and enlisted personnel. On net, cost savings in the AC are nearly $650 million per year for enlisted personnel and $254 million for officers, for a total of $902 million. This assumes that the new AC retention profile is allowed to materialize and not reshaped by policy intervention.

For RC members with prior AC service, current cost decreases by $110 million per year for enlisted personnel but increases by $13 million for officers. In the case of enlisted personnel, the decrease occurs because the number of post-20 participants, who have, on average, the highest pay, decrease by more than the increase in the number of pre-20 participants in years 10 through 20. In the case of officers, participation increases in years 12 through 25. The increase in costs associated with higher participation offsets the decrease in costs associated with the lower participation among those with fewer than 12 years. Retirement costs increase because benefits are paid for more years and start sooner than the baseline, in which members begin receiving benefits at age 60. For enlisted personnel, the increase in retirement cost is the same magnitude as the decrease in current cost, so there is no change in total cost (current cost plus retirement cost). Thus, the increase in costs for RC members with prior AC service is among officers, for whom both current cash compensation and retirement costs increase.

Table 2.3. Simulated Steady-State Army Personnel Costs (in Billions of 2013 Dollars) Under Baseline and Proposed Change

Component	Measure	Enlisted			Officer			Enlisted and Officer		
		Baseline	New	Difference	Baseline	New	Difference	Baseline	New	Difference
AC	Force size	458,220	458,220	0	90,795	90,795	0	549,015	549,015	0
	Cash compensation cost	21.750	21.600	−0.150	8.757	8.666	−0.091	30.507	30.266	−0.241
	Retirement cost	3.157	2.659	−0.498	2.187	2.024	−0.163	5.344	4.683	−0.661
	Subtotal	24.907	24.259	−0.648	10.944	10.69	−0.254	35.851	34.949	−0.902
Prior-service RC	Force size	178,491	167,882	−10,610	18,617	19,541	924	197,108	187,423	−9,686
	Cash compensation cost	1.141	1.034	−0.107	0.251	0.265	0.013	1.392	1.299	−0.094
	Retirement cost	0.093	0.206	0.113	0.037	0.094	0.057	0.130	0.300	0.170
	Subtotal	1.234	1.240	0.006	0.289	0.359	0.070	1.523	1.599	0.076
NPS RC	Force size	58,194	57,386	−808	489	489	0.000	58,683	57,875	−808
	Cash compensation cost	0.338	0.331	−0.007	0.005	0.005	0.000	0.343	0.336	−0.007
	Retirement cost	0.095	0.063	−0.032	0.001	0.001	0.000	0.096	0.064	−0.032
	Subtotal	0.433	0.394	−0.039	0.007	0.006	0.000	0.440	0.400	−0.039
Grand total		26.574	25.893	−0.681	11.240	11.055	−0.184	37.814	36.948	−0.865

NOTE: Each RC member receives 38 points per year toward retirement.

19

For RC members with no prior service, the change in participation is so slight that current cost is practically unchanged. Retirement cost decreases slightly for enlisted, by $30 million per year, because participation at years 20 and higher is somewhat lower with no change in costs for officers. In sum, costs fall by $680 million per year for enlisted and by $184 million for officers, for a total of $865 million per year. Interestingly, nearly all of the cost savings come from the AC despite the fact that there is no change in AC force size. However, prior-service RC enlisted force size decreases by more than 10,000, and, assuming that the AC-side cost savings could be identified in practice, they could be applied toward incentives to return the prior-service enlisted RC force size to its baseline level. We next describe the effects on participation and cost savings of adding RC continuation pay for enlisted personnel to restore the RC force size.

Restoring Enlisted Prior-Service Reserve Component Participation with Continuation Pay

Table 2.3 shows a fall in enlisted prior-service RC strength of 10,610 soldiers. A policy question of interest is how large of a retention bonus would be required to sustain enlisted RC strength and whether the additional bonus would fully or only partially offset the cost savings shown in Table 2.4. Here, we focus on changing retention in the prior-service RC because NPS RC retention remains essentially unchanged under the new system.

Table 2.4. Simulated Steady-State Army Personnel Costs (in Billions of 2013 Dollars) Under Baseline and Proposed Change with Continuation Pay for Enlisted Reserve Component Members with Prior Active Component Service

Component	Measure	Enlisted			Officer			Enlisted and Officer		
		Baseline	New	Difference	Baseline	New	Difference	Baseline	New	Difference
AC	Force size	458,220	458,220	0	90,795	90,795	0	549,015	549,015	0
	Cash compensation cost	21.75	21.58324	−0.15	8.757	8.666	−0.091	30.507	30.249	−0.241
	Retirement cost	3.157	2.654	−0.498	2.187	2.024	−0.163	5.344	4.678	−0.661
	Subtotal	24.907	24.237	−0.670	10.944	10.69	−0.254	35.851	34.927	−0.924
Prior-service RC	Force size	178,491	177,423	−1,068	18,617	19,541	924	197,108	196,964	−144
	Cash compensation cost	1.141	1.141	0.000	0.251	0.265	0.013	1.393	1.406	0.013
	Retirement cost	0.093	0.194	0.102	0.037	0.094	0.057	0.130	0.288	0.159
	Subtotal	1.234	1.336	0.102	0.289	0.359	0.070	1.523	1.695	0.172
NPS RC	Force size	58,194	57,386	−808	489	489	0.000	58,683	57,875	−808
	Cash compensation cost	0.338	0.331	−0.007	0.005	0.005	0.000	0.343	0.336	−0.007
	Retirement cost	0.095	0.063	−0.032	0.001	0.001	0.000	0.096	0.064	−0.032
	Subtotal	0.433	0.394	−0.039	0.007	0.006	0.000	0.440	0.400	−0.039
Grand total		26.574	25.967	−0.607	11.240	11.055	−0.184	37.813	37.022	−0.791

NOTE: Each RC member receives 38 points per year toward retirement.

The simulation capability in our model can compute optimized values of continuation payments to sustain the AC force and the RC force with prior service. Because of the flow from the AC to the RC, continuation payments to members of the RC can also affect retention in the AC. Thus, in considering what continuation payment is needed to sustain the RC force, we must also consider the effects of RC continuation pay on AC retention. We used the model's capability to identify the continuation payment for enlisted RC members with prior AC service that would be required to sustain the size of the RC force and the size and experience mix of the AC force. Note that we do not impose the constraint that the RC force experience mix must be sustained as well because the Army proposal we analyze dramatically changes the RC force shape. For this analysis, we assumed that the RC continuation payment is targeted to those with four YOS, given that we find in Figure 2.1 that enlisted RC participation decreases between years 3 and 10. The model assumes that half of the continuation pay is paid at year 4 and the remaining half is paid annually over the next three years (i.e., years 5 through 7). The continuation pay is assumed to be a multiplier times monthly basic pay at year 4. The model finds the optimal multiplier that sustains the RC force (and the AC force size and experience mix).

Figure 2.6 shows that enlisted RC participation is now higher even for those with between three and ten YOS, in contrast to Figure 2.1. It is still the case that post-20 participation is lower. The optimal multiplier is 2.91 or about three months of basic pay for an enlisted RC member with four YOS. Importantly, the change in enlisted RC force size is now 1,068 rather than 10,610, or a decrease of less than 1 percent, as seen in Table 2.4.

Figure 2.6. Simulated Steady-State Effect of the Proposal on Army Enlisted Reserve Component Participation with the Addition of Reserve Component Continuation Pay

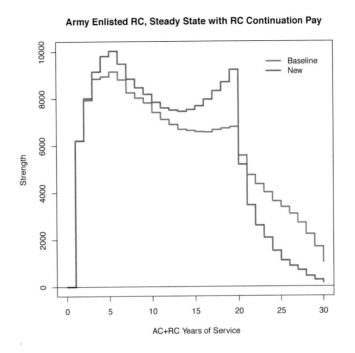

Table 2.4 replicates Table 2.3 but shows the results when continuation pay for enlisted RC personnel with prior service is included. Enlisted AC costs remain unchanged, as do NPS RC costs. As expected, prior-service RC costs increase—specifically, current cash compensation costs. Rather than a decrease in cash compensation costs of $107 million shown in Table 2.3, adding continuation pay offsets the current cash compensation cost savings. So, there is no change in prior-service RC current cash compensation costs and an overall increase in prior-service RC costs for enlisted personnel because their retirement costs increase by $102 million annually.

Despite the increase in prior-service RC costs for enlisted personnel, we still find a $607 million cost saving for enlisted personnel overall and a $791 million cost saving when officer costs are included. Thus, although the cost savings are lower when continuation pay is included, the cost savings associated with the change in experience mix in the AC force are large enough to offset the change, and, together with the cost savings associated with the officer force, the proposed change leads to an overall cost saving.

Chapter Three. Retention Results During the Transition Period

We next turn to the dynamics of change. The previous section shows the steady-state results when the entire AC and RC forces are under the new proposed system. Army planners are also concerned with how these changes unfold year by year in the transition to the steady state. The transition period reveals near- and midterm effects of the proposal. During the transition phase, the Army must decide how it will implement the policy for currently serving members. One approach is to "grandfather" current members under the current system but place members who enter the military after the policy change under the new system. By *grandfathered*, we mean that only those who enter the AC after the policy is enacted are under the new system. An alternative approach is to place current members under the new system but allow them to "opt out" in favor of the current system if they expect the value of their military career to be greater under it. In Asch, Mattock, and Hosek (2013), we extended the DRM to consider the transition to the steady state and specifically grandfathering versus opt-in/opt-out transition policies for active-duty personnel. In this study, we extended that capability to consider opt-out policies for the RC. Our extension focuses on the effects on RC participation in the transition period. We do not have RC cost estimates for the transition period; although important, that analysis must await future research. The next section shows the results of that analysis and specifically shows the effects on the RC, as well as the AC, of two implementation policies: (1) current RC members are grandfathered members under the current system, as are current AC members should they later join the RC, but new entrants are placed under the new system; and (2) current RC members are placed under the new system but have a five-year window to opt out and stay under the current system, and AC members and civilians who join the RC during a five-year window starting from the date of implementation are placed under the new system but allowed to opt out to the current system in the year that they enter the RC.

Reserve Component Transition Results

Figures 3.1 through 3.4 depict the adjustment of enlisted RC participation (among those with prior AC service) in the years after the new RC retirement system is implemented when members are grandfathered under the current system. RC members with prior AC service who join the RC after the new plan is introduced but served in the AC before the new plan was introduced would be grandfathered under the baseline current system. The figure shows enlisted RC participation for ten, 20, 30, and 40 years after the new policy begins. Comparing these figures with the results in Figure 2.1 in Chapter Two for the steady state shows that RC participation adjusts slowly, at least initially, to the new policy. Ten years after the policy has been implemented, RC participation differs only a little from the baseline steady state shown in

Figure 2.1. However, between years 10 and 20 after implementation, participation increases substantially in the midcareer. Between 20 and 40 years after implementation, participation among those with more than 20 YOS decreases, until the new steady state is reached. Interestingly, participation among those in midcareer also changes between 20 and 40 years after policy enactment as those under the new system acquire more YOS and those under the baseline system eventually leave. Figures 3.5 through 3.8 show a generally similar pattern for officers.

Figures 3.1 through 3.4 show that, in the transition and, in particular, at times $t + 20$ and $t + 30$, the projected total RC participation exceeds steady-state participation under the new compensation system. This surprising result occurs because the reservists accumulate active and reserve (AC + RC) years at different rates under the current versus the new system. RC members can start collecting retirement benefits as soon as they vest under the new system, so they have a powerful incentive to reach 20 AC + RC YOS sooner than under the current system. Under the current system, reserve retirement benefits begin generally at age 60 and do not depend on how soon individuals accrue 20 "good" years. The combination of these two sets of contrasting incentives during the transition years is that individuals "pile up" in midcareer as people under the new system flow in to the midcareer range while people under the old system are still in midcareer. Eventually, all the people under the old system flow out, and a steady state is reached.

Allowing members to opt out of the new system enables them to voluntarily stay under the current system if they expect to be better off. The advantage of allowing members to opt out is that it addresses any possible disgruntlement of currently serving members who might feel that they prefer to stay under the current system rather than move to the new system. At the same time, those who prefer the new system can continue to stay under that system. In Asch, Mattock, and Hosek (2013), we show that another advantage of permitting opting in or out is that any cost savings associated with the new system are realized sooner when existing members are given a choice of system, though we do not show the effects on costs in this analysis.

To incorporate the opt-out approach for RC members in our model, we assume that only members currently serving in the RC have the opt-out choice. We also consider different opt-out enrollment windows, where the window is the number of years after the policy is enacted when the member can choose to opt out. We consider windows that range from one year to 20 years. For example, if the opt-out window is five years, currently serving RC members are placed under the new system but have up to five years after the policy is enacted to elect to stay in the current system. If they do not choose to stay under the existing system during the five years, they are automatically under the new system.

Individuals who join the AC after the policy was enacted are automatically covered by the new system.[7] Individuals who join the RC with prior AC service during the five-year window have the choice to stay in the old system, but, if they join the RC after the five-year window, they

---

[7] Although we do not show the results for NPS reservists here, NPS RC members who joined after the policy was enacted would be automatically placed under the new system.

are automatically placed under the new system. That is, they are treated like new RC entrants and automatically covered by the new system. Thus, members serving in the AC or who are veterans with prior AC or RC service can opt into the old system only if they join the RC during the enrollment window. If they join the RC after the window, they are automatically put under the new system. For example, if the opt-in window is five years, a member who is in the AC when the new policy is enacted has up to five years to leave the AC, join the RC, and choose to opt into the old system. We show the results for the window length of five years.[8] The results for other window lengths are quite similar.

Figure 3.1. Simulated Effect of the Proposal on Army Enlisted Reserve Component Participation in the Transition to Steady State, with Grandfathered Members: Ten Years into Transition, No Choice

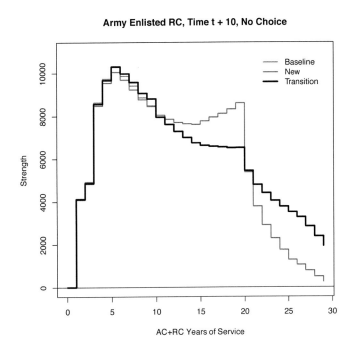

---

[8] The DRM can also incorporate the choice of when to opt in during the window. For example, if the window is five years, a member might choose to defer the opt-in decision until more information is available, or the member might choose to opt in immediately. In other analysis for the AC, we have modeled this choice of when during the window to opt in. We do not include that choice in the analysis presented here.

Figure 3.2. Simulated Effect of the Proposal on Army Enlisted Reserve Component Participation in the Transition to Steady State, with Grandfathered Members: 20 Years into Transition, No Choice

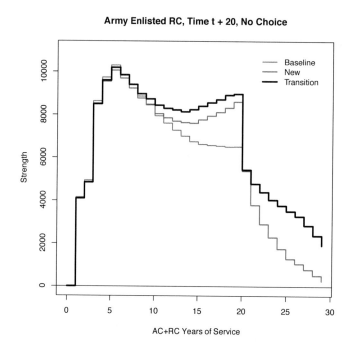

Figure 3.3. Simulated Effect of the Proposal on Army Enlisted Reserve Component Participation in the Transition to Steady State, with Grandfathered Members: 30 Years into Transition, No Choice

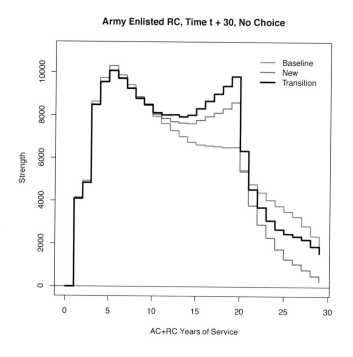

Figure 3.4. Simulated Effect of the Proposal on Army Enlisted Reserve Component Participation in the Transition to Steady State, with Grandfathered Members: 40 Years into Transition, No Choice

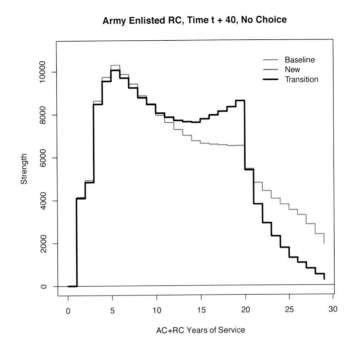

Figure 3.5. Simulated Effect of the Proposal on Army Officer Reserve Component Participation in the Transition to Steady State, with Grandfathered Members: Ten Years into Transition, No Choice

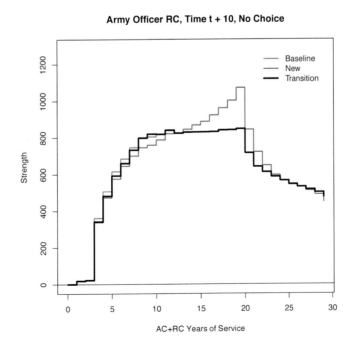

Figure 3.6. Simulated Effect of the Proposal on Army Officer Reserve Component Participation in the Transition to Steady State, with Grandfathered Members: 20 Years into Transition, No Choice

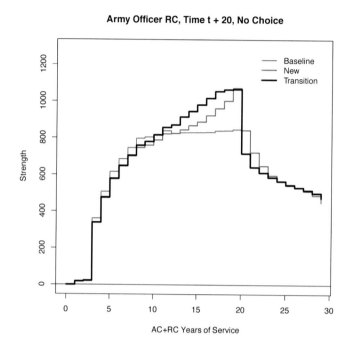

Army Officer RC, Time t + 20, No Choice

Figure 3.7. Simulated Effect of the Proposal on Army Officer Reserve Component Participation in the Transition to Steady State, with Grandfathered Members: 30 Years into Transition, No Choice

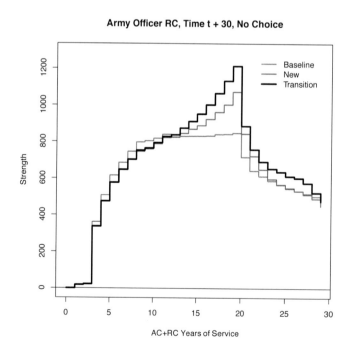

Army Officer RC, Time t + 30, No Choice

Figure 3.8. Simulated Effect of the Proposal on Army Officer Reserve Component Participation in the Transition to Steady State, with Grandfathered Members: 40 Years into Transition, No Choice

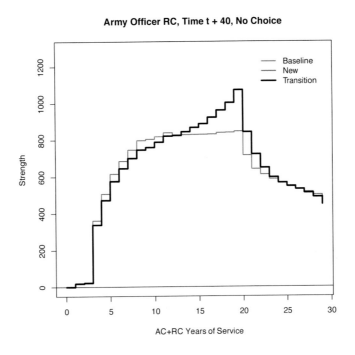

Figures 3.9 through 3.12 depict the adjustment of enlisted RC participation (among those with prior AC service) in the years after the new RC retirement system is implemented when members are given the choice of opting out to the old system during a five-year enrollment window. The figures show enlisted RC participation for five, 15, 25, and 35 years after the new policy begins, depicted by the blue line in each chart. The red line in each chart shows RC participation among those who choose to remain under the current system. The difference between the blue and red lines is the participation of those who are under the new system.

Three results are apparent from Figures 3.9 through 3.12. First, comparing the results with the steady-state results in Figure 2.1 in Chapter Two, we see that RC participation adjusts quickly to the new policy, unlike the grandfathering case in Figures 3.1 through 3.4. By year 5, when the five-year window is ended, much of the change in participation has occurred. Second, most RC participants stay in the new system, as seen by the relatively low level of the red line five years after the policy has begun. It is not surprising that the new steady state is reached quickly given that so many RC members stay in the new system, i.e., few opt for the old system. Finally, the decision to enter the old system is quite uniform across years of active and reserve service, though there is some drop after 20 YOS for enlisted personnel. That is, in year 5, the red line is relatively flat across YOS, implying that the number choosing the old system is quite similar, regardless of YOS. As time elapses, members who were serving at the time of the policy change eventually leave service, and only new members who are automatically covered by the new system make up the bulk of the RC force. In the new steady state, no member is under the current system.

Despite the uniformity across YOS in the number choosing the old system, there are differences in this decision by age. Figures 3.13 through 3.16 show RC participation among those with prior AC service by age rather than by years of active and reserve service. These figures correspond to the results shown in Figure 2.2 in Chapter Two. As in Figures 3.9 through 3.12, the blue line in Figures 3.13 through 3.6 shows total participation in each transition year, assuming a five-year window, while the red line shows participation among those choosing the old (baseline) system. Here we see that older RC members are more likely to opt to stay under the old system, while younger members are more likely to be covered by the new system. This result is unsurprising—members who are close to the current retirement eligibility age of 60 are unlikely to find much benefit to switching to an immediate annuity and so are likely to choose to stay under the old system. As time elapses, members who opted to stay under the old system leave the RC and RC strength is made up more and more of individuals who entered the military after the policy was enacted and were automatically placed under the new system. Thus, as time elapses, fewer RC members are under the old system.

Figure 3.9. Simulated Effect of the Proposal on Army Enlisted Reserve Component Participation in the Transition to Steady State, with Opt-Out Choice During Five-Year Window: Five Years into Transition

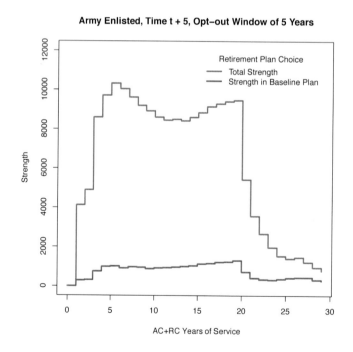

32

Figure 3.10. Simulated Effect of the Proposal on Army Enlisted Reserve Component Participation in the Transition to Steady State, with Opt-Out Choice During Five-Year Window: 15 Years into Transition

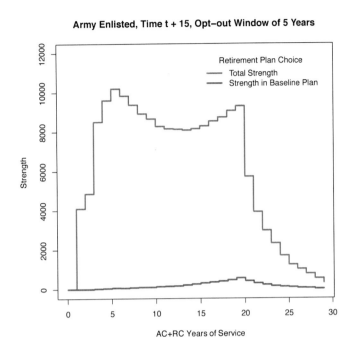

Figure 3.11. Simulated Effect of the Proposal on Army Enlisted Reserve Component Participation in the Transition to Steady State, with Opt-Out Choice During Five-Year Window: 25 Years into Transition

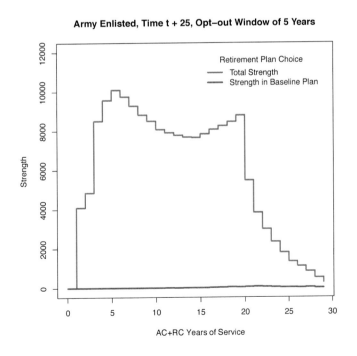

33

Figure 3.12. Simulated Effect of the Proposal on Army Enlisted Reserve Component Participation in the Transition to Steady State, with Opt-Out Choice During Five-Year Window: 35 Years into Transition

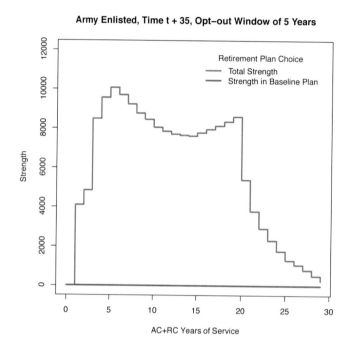

Figure 3.13. Simulated Effect of the Proposal on Army Enlisted Reserve Component Participation by Age in the Transition to Steady State, with Opt-Out Choice During Five-Year Window: Five Years into Transition

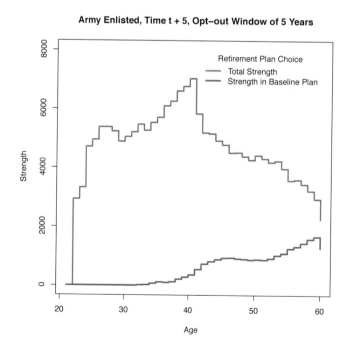

Figure 3.14. Simulated Effect of the Proposal on Army Enlisted Reserve Component Participation by Age in the Transition to Steady State, with Opt-Out Choice During Five-Year Window: 15 Years into Transition

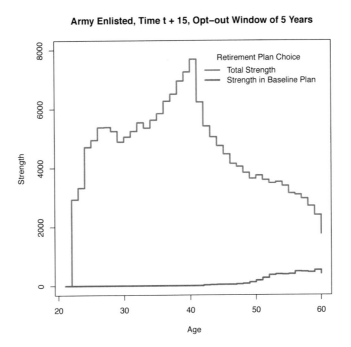

Figure 3.15. Simulated Effect of the Proposal on Army Enlisted Reserve Component Participation by Age in the Transition to Steady State, with Opt-Out Choice During Five-Year Window: 25 Years into Transition

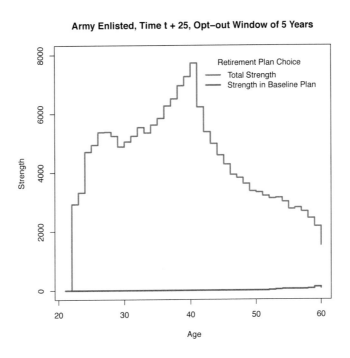

Figure 3.16. Simulated Effect of the Proposal on Army Enlisted Reserve Component Participation by Age in the Transition to Steady State, with Opt-Out Choice During Five-Year Window: 35 Years into Transition

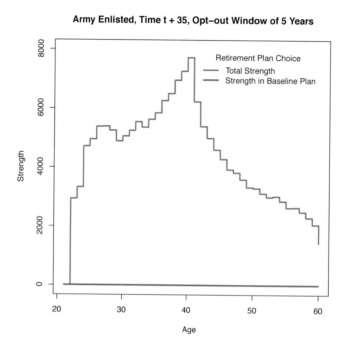

Figures 3.17 through 3.20 show the results for officers where the x-axis measures active and reserve YOS. These figures correspond to Figures 3.9 through 3.12 for enlisted personnel. Figures 3.21 through 3.24 show the officer results when the x-axis measures age, corresponding to Figures 3.13 through 3.16 for enlisted personnel. The results for officers are quite similar to those for enlisted. As with enlisted, the new steady state is reached far more quickly when RC members are defaulted into the new system (or opt out and stay under the old system), relatively few choose to stay under the old system, and those who choose to stay under the old system tend to be older RC members but differ little in terms of their years of active and reserve service.

Figure 3.17. Simulated Effect of the Proposal on Army Officer Reserve Component Participation in the Transition to Steady State, with Opt-Out Choice During Five-Year Window: Five Years into Transition

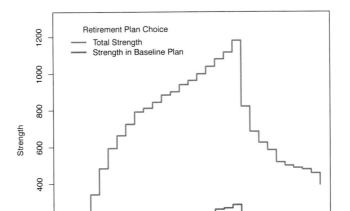

Figure 3.18. Simulated Effect of the Proposal on Army Officer Reserve Component Participation in the Transition to Steady State, with Opt-Out Choice During Five-Year Window: 15 Years into Transition

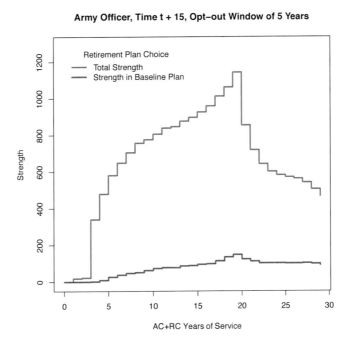

Figure 3.19. Simulated Effect of the Proposal on Army Officer Reserve Component Participation in the Transition to Steady State, with Opt-Out Choice During Five-Year Window: 25 Years into Transition

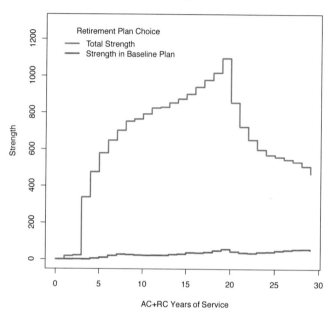

Figure 3.20. Simulated Effect of the Proposal on Army Officer Reserve Component Participation in the Transition to Steady State, with Opt-Out Choice During Five-Year Window: 35 Years into Transition

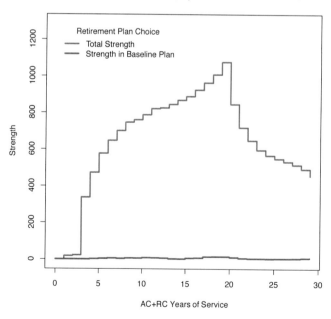

Figure 3.21. Simulated Effect of the Proposal on Army Officer Reserve Component Participation by Age in the Transition to Steady State, with Opt-Out Choice During Five-Year Window: Five Years into Transition

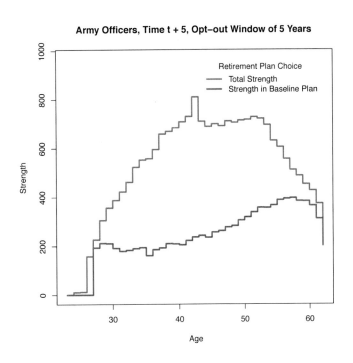

Figure 3.22. Simulated Effect of the Proposal on Army Officer Reserve Component Participation by Age in the Transition to Steady State, with Opt-Out Choice During Five-Year Window: 15 Years into Transition

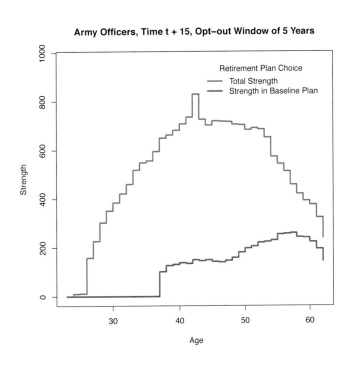

Figure 3.23. Simulated Effect of the Proposal on Army Officer Reserve Component Participation by Age in the Transition to Steady State, with Opt-Out Choice During Five-Year Window: 25 Years into Transition

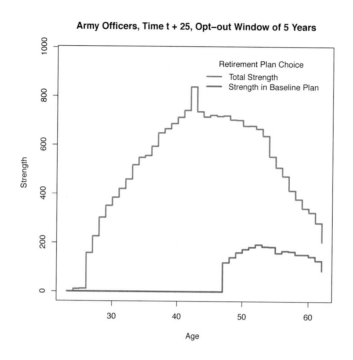

Figure 3.24. Simulated Effect of the Proposal on Army Officer Reserve Component Participation by Age in the Transition to Steady State, with Opt-Out Choice During Five-Year Window: 35 Years into Transition

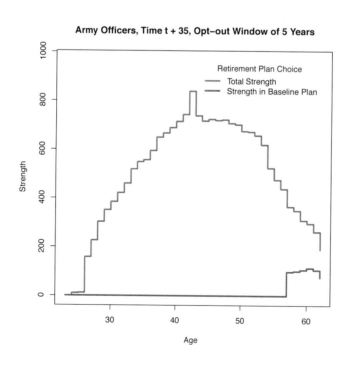

Active Component Transition Results

As we showed in Figure 2.3 in Chapter Two, the change in the RC retirement system has a small but noticeable effect on AC retention. We found AC retention slightly higher among junior AC personnel as RC (and therefore AC) service becomes more valuable, but it is also lower in years just prior to and just after retirement vesting because soldiers nearing and just after 20 YOS give up little in retirement benefits by switching to the RC, even late in their midcareers. As mentioned, one might make such a switch if he or she has a particularly good external opportunity. Because of these effects on the AC in the steady state, we would expect AC retention to change during the transition period. We show those results here.

Consistently with the results for the RC, grandfathering current RC members in the current system, with choice to enter the new system, leads to a change in AC retention that occurs only slowly over time. These results are not shown, but, like that in the RC, AC retention changes little in the first ten years and then changes more after ten years of elapsed time since the policy change, when more members are covered by the new RC retirement system.

Figures 3.25 through 3.28 and 3.29 through 3.32 show, for enlisted personnel and for officers, respectively, that depict the adjustment of AC retention in the years after the new RC retirement system is implemented and assuming that members have a five-year window in which to opt out of the new system. Again, this involves leaving the AC and joining the RC within the five-year window, in which case a member would be able to choose between the current and the new retirement system. In particular, AC retention is shown five, 15, 25, and 35 years after the new policy begins. Comparing these charts with the results in Figure 2.3 for the steady state shows that AC retention for both enlisted personnel and officers adjusts quickly to the new policy. By year 5 (i.e., five years after the policy has been implemented), much of the change in retention has occurred, with further changes occurring in later years as members under the old system flow out of the military and those under the new system accumulate more YOS.

Figure 3.25. Simulated Effect of the Proposal on Army Enlisted Active Component Retention in the Transition to Steady State, with Opt-Out Choice During Five-Year Window: Five Years into Transition

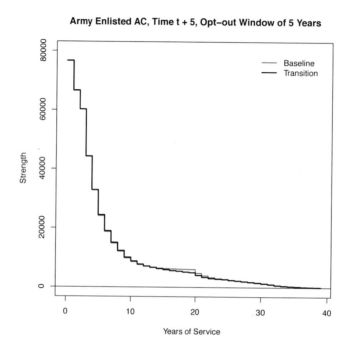

Army Enlisted AC, Time t + 5, Opt–out Window of 5 Years

Figure 3.26. Simulated Effect of the Proposal on Army Enlisted Active Component Retention in the Transition to Steady State, with Opt-Out Choice During Five-Year Window: 15 Years into Transition

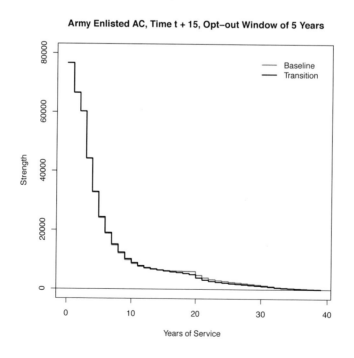

Army Enlisted AC, Time t + 15, Opt–out Window of 5 Years

Figure 3.27. Simulated Effect of the Proposal on Army Enlisted Active Component Retention in the Transition to Steady State, with Opt-Out Choice During Five-Year Window: 25 Years into Transition

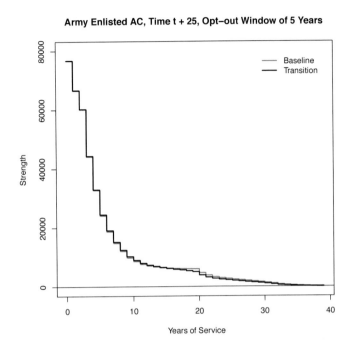

Figure 3.28. Simulated Effect of the Proposal on Army Enlisted Active Component Retention in the Transition to Steady State, with Opt-Out Choice During Five-Year Window: 35 Years into Transition

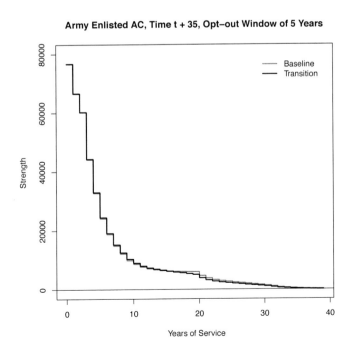

Figure 3.29. Simulated Effect of the Proposal on Army Officer Active Component Retention in the Transition to Steady State, with Opt-Out Choice During Five-Year Window: Five Years into Transition

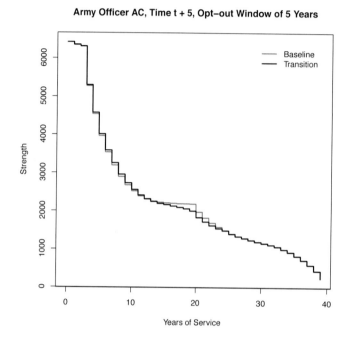

Army Officer AC, Time t + 5, Opt–out Window of 5 Years

Figure 3.30. Simulated Effect of the Proposal on Army Officer Active Component Retention in the Transition to Steady State, with Opt-Out Choice During Five-Year Window: 15 Years into Transition

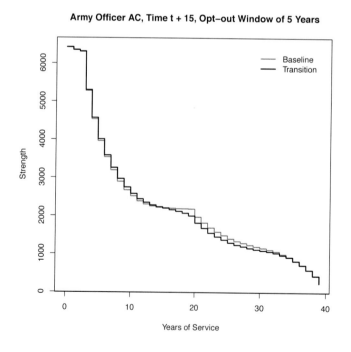

Army Officer AC, Time t + 15, Opt–out Window of 5 Years

Figure 3.31. Simulated Effect of the Proposal on Army Officer Active Component Retention in the Transition to Steady State, with Opt-Out Choice During Five-Year Window: 25 Years into Transition

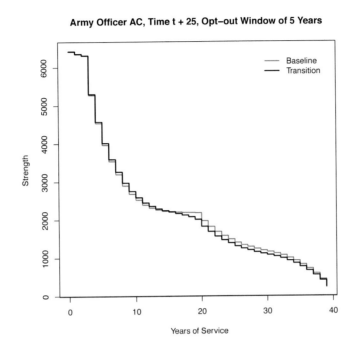

Figure 3.32. Simulated Effect of the Proposal on Army Officer Active Component Retention in the Transition to Steady State, with Opt-Out Choice During Five-Year Window: 35 Years into Transition

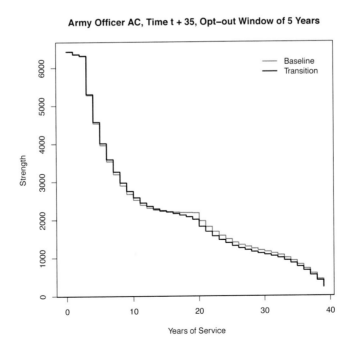

Chapter Four. Conclusion

The results of the analysis reveal that a reserve retirement benefit system in which retirement benefits began immediately upon retirement from the Selected Reserve, and for which retirement points accumulated at a rate of one point per day of reserve service, whether on inactive duty or active duty, would be feasible and cost-effective relative to the current reserve retirement benefit system. However, it would result in a different experience mix for both the AC and RC, but especially the RC. It would sustain the AC and RC force sizes, although this would also require a supplement to current RC compensation, and it would cost less than the current system. These findings are based on applying the proposed system to the Army, but we expect that the results for other services would be similar. The estimated cost savings for the Army were about $800 million per year. Also, we note that the analysis assumed that the AC retirement benefit system was in its current form, as was the formula for computing an individual's retirement benefit. Changes to the AC retirement benefit system or the benefit formula could alter the results; however, our model could analyze the RC proposal under different possible changes in the AC system or the formula.

In addition to the above results, the proposal would affect prior-service RC participation and AC retention. But there was practically no effect on NPS RC participation.

The change in prior-service RC participation would make the RC force less senior, as participation increased for enlisted and officers in the midcareer years up to 20 YOS, where *years* includes both AC and RC years. Enlisted participation decreased after 20 YOS, and officer participation was higher until the 25th YOS and decreased afterward. Still, the net effect would be to make both the RC enlisted and officer forces less senior.

AC retention would increase in YOS 5 to 15 and decrease in the years before and after 20 YOS, which is when retirement benefits vest. The increase occurs because AC members recognize that they might join the RC in the future, and if they did, the combination of more years in the AC and the availability of retirement benefits in the RC immediately upon retirement from the RC—rather than at age 60, as under the current system—increases the value of staying in the AC over these midcareer years. But AC retention is lower around 20 YOS because an AC members gives up little by leaving and joining the RC, again because RC retirement benefits begin immediately upon RC retirement. The decrease in AC retention just prior to and after 20 YOS was the major factor in cost savings. This decrease in senior personnel led to decreases in the cost of current compensation and the cost of retirement benefits. (The retirement benefit cost is an accrual charge representing funds that should be set aside currently to fund outlays for future retirement benefits.)

Another aspect of the proposal relevant to the AC is that force planners may be more willing to shape the AC force and achieve shorter or longer AC careers knowing that AC members who are separated will be eligible for immediate retirement benefits if they join the RC and vest in the RC. That is, planners may be willing to shape the AC without concern about denying retirement benefits to those who are not yet vested for AC benefits. Thus, the proposal has the potential to enhance AC force management flexibility.

The transition analysis showed the year-by-year changes in force shape resulting from individual decisions to participate in the RC and to stay in or leave the AC. Importantly, we found that a transition policy in which RC members were given a choice between the two systems sped up the transition to the steady state. Offering this choice is also in the individual's interest. The individual can be no worse off, and often better off, by having the opportunity to choose the system that he or she prefers. Thus, providing members with an immediate option to transition to the new system has the potential to make service members better off *and* lower costs.

Appendix. Simulation Tables

This appendix contains tables showing the simulated change over time in Army strength when reserve members who are grandfathered under the current system can switch. Table A.1 shows count data for officers and enlisted by YOS categories, and Table A.2 shows percentages by YOS categories.

Table A.1. Army Strength Within Year-of-Service Groups by Years Elapsed Since Policy Change

Personnel	0	5	15	25	35	41
AC officers						
1 to 5	28,971	29,028	29,033	29,033	29,033	29,033
6 to 10	16,318	16,555	16,623	16,623	16,623	16,623
11 to 15	11,691	11,732	11,929	11,930	11,930	11,930
16 to 20	10,917	10,469	10,574	10,598	10,598	10,598
21+	22,898	22,673	21,783	21,459	21,468	21,471
Prior-service RC officers						
1 to 5	906	870	857	854	851	853
6 to 10	3,661	3,612	3,471	3,444	3,437	3,435
11 to 15	4,302	4,676	4,360	4,222	4,153	4,157
16 to 20	4,442	5,651	5,210	5,081	4,915	4,857
21+	5,876	5,682	6,148	6,073	6,019	5,918
AC enlisted						
1 to 5	280,268	280,812	280,812	280,812	280,812	280,812
6 to 10	79,378	80,896	81,317	81,317	81,317	81,317
11 to 15	36,155	36,323	37,427	37,427	37,427	37,427
16 to 20	30,382	27,285	27,806	27,982	27,982	27,982
21+	32,038	30,995	26,687	25,889	25,920	25,920
Prior-service AC enlisted						
1 to 5	27,446	27,331	27,093	27,001	26,986	26,955
6 to 10	46,934	48,107	47,087	46,435	46,221	46,205
11 to 15	36,592	42,341	41,000	39,175	38,869	38,869
16 to 20	32,774	45,055	44,204	41,536	40,824	40,778
21+	36,392	22,509	20,370	20,623	20,078	20,017

Table A.2. Army Percentage Within Year-of-Service Groups by Years Elapsed Since Policy Change

Personnel	0	5	15	25	35	41
AC officers						
1 to 5	31.9	32.1	32.3	32.4	32.4	32.4
6 to 10	18.0	18.3	18.5	18.5	18.5	18.5
11 to 15	12.9	13.0	13.3	13.3	13.3	13.3
16 to 20	12.0	11.6	11.8	11.8	11.8	11.8
21+	25.2	25.1	24.2	23.9	23.9	23.9
Prior-service RC officers						
1 to 5	4.7	4.2	4.3	4.3	4.4	4.4
6 to 10	19.1	17.6	17.3	17.5	17.7	17.9
11 to 15	22.4	22.8	21.7	21.5	21.4	21.6
16 to 20	23.2	27.6	26.0	25.8	25.4	25.3
21+	30.6	27.7	30.7	30.9	31.1	30.8
AC enlisted						
1 to 5	61.2	61.5	61.8	61.9	61.9	61.9
6 to 10	17.3	17.7	17.9	17.9	17.9	17.9
11 to 15	7.9	8.0	8.2	8.3	8.3	8.3
16 to 20	6.6	6.0	6.1	6.2	6.2	6.2
21+	7.0	6.8	5.9	5.7	5.7	5.7
Prior-service AC enlisted						
1 to 5	15.2	14.7	15.1	15.4	15.6	15.6
6 to 10	26.1	26.0	26.2	26.6	26.7	26.7
11 to 15	20.3	22.8	22.8	22.4	22.5	22.5
16 to 20	18.2	24.3	24.6	23.8	23.6	23.6
21+	20.2	12.1	11.3	11.8	11.6	11.6

References

Asch, Beth J., James Hosek, and Michael G. Mattock, *A Policy Analysis of Reserve Retirement Reform*, Santa Monica, Calif.: RAND Corporation, MG-378-OSD, 2013. As of December 10, 2013:
http://www.rand.org/pubs/monographs/MG378.html

Asch, Beth J., Michael G. Mattock, and James Hosek, *A New Tool for Assessing Workforce Management Policies Over Time: Extending the Dynamic Retention Model*, Santa Monica, Calif.: RAND Corporation, RR-113-OSD, 2013. As of December 10, 2013:
http://www.rand.org/pubs/research_reports/RR113.html

Goldberg, Matthew S., *A Survey of Enlisted Retention: Models and Findings*, Alexandria, Va.: CNA, CRM D0004085.A2/Final, November 2001. As of December 10, 2013:
http://www.cna.org/sites/default/files/research/D0004085.A2.pdf

Gotz, Glenn A., "Comment on 'The Dynamics of Job Separation: The Case of Federal Employees,'" *Journal of Applied Econometrics*, Vol. 5, No. 3, July–September 1990, pp. 263–268.

Hosek, James, and Trey Miller, *Effects of Bonuses on Active Component Reenlistment Versus Prior Service Enlistment in the Selected Reserve*, Santa Monica, Calif.: RAND Corporation, MG-1057-OSD, 2011. As of December 10, 2013:
http://www.rand.org/pubs/monographs/MG1057.html

Mattock, Michael G., James Hosek, and Beth J. Asch, *Reserve Participation and Cost Under a New Approach to Reserve Compensation*, Santa Monica, Calif.: RAND Corporation, MG-1153-OSD, 2012. As of December 10, 2013:
http://www.rand.org/pubs/monographs/MG1153.html

U.S. Department of Defense, *Report of the Eleventh Quadrennial Review of Military Compensation: Main Report*, Washington, D.C., June 2012. As of December 10, 2013:
http://militarypay.defense.gov/reports/qrmc/11th_QRMC_Main_Report_(290pp)_Linked.pdf